This book belongs to
Hector's best friend

....

....

To my beautiful granddaughter, Maya, who loved Hector from the very first story.

First published in 2026 by Hector Books
Perth, Western Australia WA 6111

2025 Illustrations by Mario Clemente and Gerry Gillougley
Edited by Anne Orton

A catalogue record for this book is available from the National Library of Australia

ISBN: 9781763610521 - Hardback
ISBN: 9781763610538 - Paperback

Printed globally by Ingram Spark

www.hectorbooks.com.au

Hector, Hamish and Morag

Life on the Farm

Christine Gillougley

Hector woke up early, the sun was shining and it was already warm. He jumped up and shook himself. Hairy fur flew everywhere. 'ATCHOO' he sneezed.

He made his way over to the fresh green grass, munching slowly as he decided what to do now his two friends had arrived.

Every school day children came to see the cow family. Hector loved it as most had never seen a highland cow before, let alone one as cute as Hector.

He would stand as one by one each child came up, patted him, spoke to him, then cuddled him. That was the best part.

He even gave them rides on his back, now Hamish and Morag would be part of this too.

It was a Sunday and Hector had so much to show Hamish and Morag that he couldn't contain himself. 'Time to get up' he shouted, 'rise and shine everyone.'

One by one they stood up and stretched. 'What a lovely sleep that was' said Hamish. 'I wasn't cold at all' he laughed.

Hector showed them the best grass and told them to get munching as they were heading out on an adventure.
'How exciting' said Morag.

Once their stomachs were full, Hector took them to meet some of his friends on the farm.

'Every morning' said Hector, 'Mrs. Maclean goes into the hen house and collects the freshly laid eggs and takes them up to the farm for breakfast.'

'Here she comes now' said Morag, 'let's go and help. 'We can carry them for her.'

As they looked inside Hector said. 'That is Camilla, she is the best layer on the farm.' The three little cows laughed, as the chicks ran around the coop.

'Let's use this bag and we can stand at the gate and gather the eggs.'

'Sounds like a good idea', said Mrs Maclean 'come on, I will go back inside and pass the eggs to Morag. Be careful not to break them.'

Mrs Maclean went inside the hen house. After a few minutes she came out. 'Here you go Morag, Camilla and the girls have been busy last night.'

She was covered in straw. 'Ha ha' they all laughted as she dusted herself down, 'you look like us. All hairy.'

'Let's all head up to the house,' said Mrs.Maclean. 'Donald will be needing his breakfast.'

Hamish, carrying the bag around his horns said, 'Hope you like your eggs scrambled.' They all laughed when they saw Donald coming in the kitchen door.

'Thank you for your help' said Mrs. Maclean, as she went inside to start cooking breakfast.

'You're welcome' said the three cows together. 'See you later.'

They spent the day meeting the gang. Piggy the pig shared all the farm's do's and don'ts.

In the afternoon, they walked to the duck pond.

Daphne the duck paddled in the pond, keeping cool. The little cows jumped in and chased the ducks. They laughed and played until it was time to head back.

The next morning, the three little cows continued their journey around the farm. They spotted Missy the dog sitting in a field with one paw off the ground, while her two pups, Dougal and Rory, barked and ran in circles.

'Let's go help!' said Hamish, and they all ran over.

'What's wrong?' asked Morag. 'Did you hurt your leg?' 'Yes,' said Missy. 'I can't put it down, and the pups are all ready for puppy training!'

'Don't worry,' said Hamish. 'You rest your leg, we'll take Dougal and Rory to class.'

The three little cows headed into town with the pups and joined the class at the local park. They lined up with all the puppies. What a funny sight it was.

'Sit!' called the teacher. Everyone sat.
'Up!' she shouted. They all stood.
'Come here!' They all trotted over.

'Well done' said the teacher. 'Now it's time for some fun.' She pulled out a big, long tunnel.

'You will have to be brave to run through this!'

The lead dog stood at the tunnel's entrance, then disappeared inside and popped out the other side. Everyone clapped and cheered. 'Rory, you go first.'

Rory stepped up, took a big deep breath, ran in and popped out the other side.

'Well done!' everyone clapped. 'Dougal, you're next.' He felt nervous, but knowing his brother had done well gave him courage.

He took a breath, ran in one side and popped out the other. Everyone cheered again.

Hector was next. It was a tight squeeze, but in he went and **PoP!** - out he came. **'Hoorah!'** shouted everyone.

Hamish was scared of the dark, but he knew he had to try. He took a deep breath, stepped into the tunnel and slowly began crawling along.

Suddenly, he got stuck, right in the middle. He couldn't move forward or back.

'Help!' he cried. 'I'm well and truly stuck!'

'Come on Morag, let's help' said Hector. Hector crawled in from the front, and Morag from the back. **'Push!'** said Hector. **'Pull!'** said Morag.

'One – two – three – **POP!**'

Hector, Hamish and Morag all tumbled out, rolling head over heels onto the grass.

'**Ta-da!**' laughed Hector. 'That's how it's done!'

After everyone had recovered from giggling, it was time to head home.

'That was great fun' they all said.
'We'll have to come back another day.'

'Come on,' said Morag, 'time to head back to the farm and check on Mum.' They made their way along the path, staying close together as the sky grew darker.

When they got back, Missy was feeling much better and thanked the three little cows for helping out.

Dougal and Rory went straight to bed, while Hector, Hamish and Morag wandered back to the barn, stopping to nibble some grass.

They lay down together and chatted with Hector's mum until, one by one, they all drifted off to sleep.

What an exciting day it had been. Hamish and Morag had made new friends, just by being helpful and kind.

'I wonder what tomorrow will bring?' thought Hector's mum, as she gently closed the barn door.

The Adventures of Hector Hamish and Morag

the complete collection of books in the series

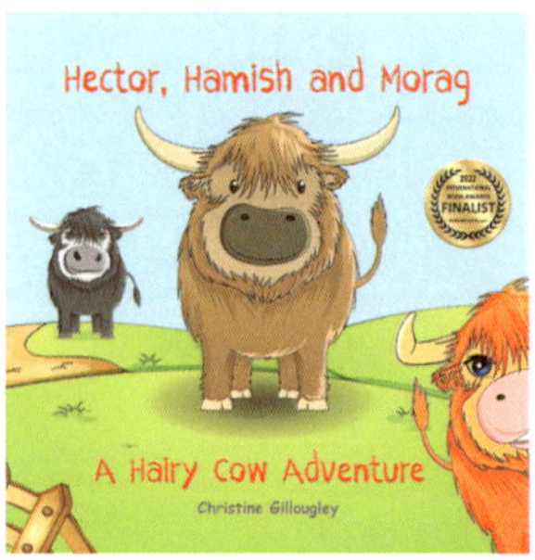

A Hairy Cow Adventure

Life on the Farm

The Beach Rescue

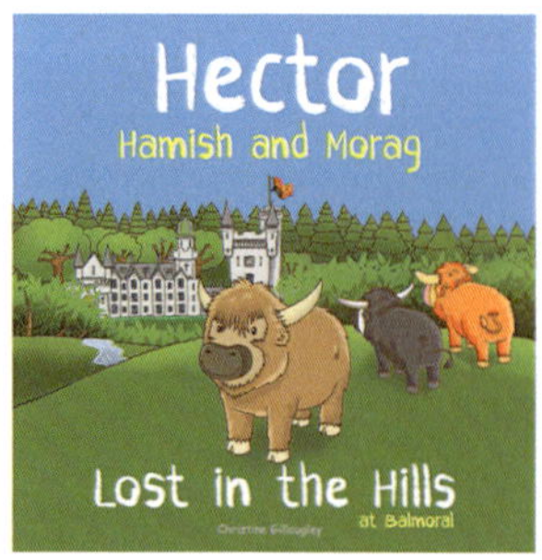

Lost in the Hills

www.ingramcontent.com/pod-product-compliance
Lightning Source LLC
LaVergne TN
LVRC101545220926
848486LV00041B/379

9781763610521